Andrew Morgan

**Flora of the Miami Valley, Ohio**

Andrew Morgan

**Flora of the Miami Valley, Ohio**

ISBN/EAN: 9783337268657

Printed in Europe, USA, Canada, Australia, Japan

Cover: Foto ©berggeist007 / pixelio.de

More available books at **www.hansebooks.com**

OF THE

# MIAMI VALLEY, OHIO.

— BY —

## A. P. MORGAN.

PUBLISHED BY THE

LITERARY UNION,

DAYTON, OHIO.

1878.

# PREFACE.

This Fiora of the Miami Valley, I have prepared for the Botanical Section of the Literary Union, at the suggestion of Hon. A. D. Wilt, President of the Society. It embodies the results of rambles after plants, during many summers, in various parts of the valley of the Miami.

Never having in view presenting my researches in systematic shape or putting them in permanent form, I did not aim, at any time, to make thorough and exhaustive work. My studies of plants have been pursued as a delightful recreation—as a relaxation from toil and business. No one, then, more than myself, can be aware of the deficiencies of the catalogue of plants here presented. Several of the Natural Orders are little worked up. This is plain in reference to Grasses and Sedges; it is also true in reference to the aquatic plants of most of the orders. It is hoped that students of Botany in every part of the region will feel stimulated to look for plants not found in the catalogue, in order to make up these deficiencies in a future edition.

I am indebted to Mrs. E. J. Spence, of Springfield, O., for a list of the Ferns, Mosses, Hepaticæ, and Lichens which she has collected in that locality.

I have made use, also, of a catalogue of the trees growing in Woodland Cemetery, prepared many years ago by MR. JOHN W. VAN CLEVE.

Besides its general scientific value, such a catalogue of the plants of the Miami Valley will, of course, be exceedingly useful and convenient to all resident botanists who make exchanges and accumulate herbaria. It is thought, also, that it may prove of service to pupils in the High Schools and other institutions of learning while pursuing the usual course of instruction in Botany.

It is my desire, at an early day, to issue a revised edition of this Flora, which shall be complete and exhaustive. In it, also, I propose to indicate the *locality* of species which are rare and not common to the whole region. That this may be done in the best manner possible, all lovers and students of Botany living in every part of the Miami Valley, are requested to furnish me with lists of additional species, and with such other information as may be available for this purpose.

A. P. MORGAN.

Dayton, Ohio, April 10, 1878.

# FLORA

# MIAMI VALLEY, OHIO.

----

*SERIES I.*

PHÆNOGAMOUS OR FLOWERING PLANTS.

*CLASS I.*

DICOTYLEDONOUS OR EXOGENOUS PLANTS

SUB-CLASS *I.*—ANGIOSPERMÆ.

DIVISION *I.*—POLYPETALOUS.

----

**RANUNCULACEÆ** — The Buttercup Family.

CLEMATIS, L.

C. **Viorna**, L.   Leather-flower.
C. **Virginiana**, L.   Virgin's Bower.

ANEMONE, L.

A. **Virginiana**, L.   Virginian Anemone.
A. **Pennsylvanica**, L.   Pennsylvanian Anemone.

## HEPATICA, Dill.

H. acutiloba, DC.   Acute-lobed Hepatica.

## THALICTRUM. Tourn.

T. anemonoides, Michx.   Rue Anemone.
T. dioicum, L.   Early Meadow Rue.
T. Cornuti, L.   Tall Meadow Rue.

## RANUNCULUS, L.

R. divaricatus, Schrank.   White Water Buttercup.
R. aquatilis, L.
var. trichophyllus, Chaix.   Common White Water
    Buttercup.
R. multifidus, Pursh.   Yellow Water Buttercup.
R. abortivus, L.   Small-flowered Buttercup.
var. micranthus, Gray.   Hairy Small-flowered But-
    tercup.
R. sceleratus, L.   Cursed Buttercup.
R. recurvatus, Poir.   Hooked Buttercup.
R. repens, L.   Creeping Buttercup.

## CALTHA, L.

C. palustris, L.   Marsh Marigold.

## AQUILEGIA, Tourn.

A. Canadensis, L.   Wild Columbine.

## DELPHINIUM, Tourn.

D. tricorne, Michx.   Dwarf Larkspur.
D. consolida, L.   Field Larkspur.

## HYDRASTIS, L.

H. Canadensis, L.   Yellow Puccoon.

ACTÆA, L.

A. spicata, L.
var. rubra, Michx. Red Baneberry.
A. alba, Bigel. White Baneberry.

CIMICIFUGA, L.

C. racemosa, Ell. Black Snakeroot.

---

# MAGNOLIACEÆ—The Magnolia Family.

LIRIODENDRON, L.

L. Tulipifera, L. Tulip Tree.

# ANONACEÆ—The Custard Apple Family.

ASIMINA, Adans.

A. triloba, Dunal. Papaw.

---

# MENISPERMACEÆ — The Moonseed Family.

MENISPERMUM, L.

M. Canadense, L. Canadian Moonseed.

---

# BERBERIDACEÆ—The Barberry Family.

BERBERIS, L.

B. vulgaris, L. Common Barberry.

CAULOPHYLLUM, Michx.

C. thalictroides, Michx. Blue Cohosh.

JEFFERSONIA, Barton.

J. **diphylla,** Pers.  Twin-leaf.

PODOPHYLLUM, L.

P. **peltatum,** L.  May Apple.

---

**NYMPHÆACEÆ** — The Water Lily Family.

NUPHAR. Smith.

N. **advena,** Ait.  Yellow Pond Lily.

---

**PAPAVERACEÆ** — The Poppy Family.

PAPAVER, L.

P. **somniferum,** L.  Common Poppy.

ARGEMONE, L.

A. **Mexicana,** L.  Mexican Poppy.

CHELIDONIUM, L.

C. **majus,** L.  Celandine.

SANGUINARIA, Dill.

S. **Canadensis,** L.  Bloodroot.

---

**FUMARIACEÆ** — The Fumitory Family.

ADLUMIA, Raf.

A. **cirrhosa,** Raf.  Climbing Fumitory.

DICENTRA, Bork.

D. **cucullaria,** DC.  Dutchman's Breeches.

D. **Canadensis,** DC.  Squirrel Corn.

CORYDALIS. Vent.

C. flavula, Raf. Yellow Corydalis.

FUMARIA. L.

F. officinalis, L. Common Fumitory.

CRUCIFERÆ—The Mustard Family.

NASTURTIUM. R. Br.

N. officinale, R. Br. Water Cress.
N. sessiliflorum, Nutt.
N. palustre, DC. Marsh Cress.
N. Armoracia, Fries. Horse Radish.

DENTARIA. L.

D. laciniata, Muhl. Pepper-root.

CARDAMINE. L.

C. rhomboidea, DC. Spring Cress.
var. purpurea, Torr. Purple Spring Cress.
C. hirsuta, L. Small Bitter Cress.
var. sylvatica, Gray. Wood Cress.

ARABIS. L.

A. dentata, Torr. & Gray. Rock Cress.
A. hirsuta, Scop. Common Rock Cress.
A. lævigata, DC. Smooth Rock Cress.
A. Canadensis, L. Sickle-pod.
A. hesperidoides, Gray. False Rocket.

BARBAREA. R. Br.

B. vulgaris, R. Br. Yellow Rocket.

SISYMBRIUM. L.

S. officinale, Scop.  Hedge Mustard.
L. canescens, Nutt.  Tansy Mustard.

BRASSICA. TOURN.

B. nigra, Gray.  Black Mustard.

CAMELINA. CRANTZ.

C. sativa, Crantz.  False Flax.

CAPSELLA. VENT.

C. Bursa-pastoris.  Mœnch.

LEPIDIUM. L.

L. Virginicum, L.  Wild Peppergrass.

RAPHANUS. L.

R. Raphanistrum, L.  Wild Radish.
R. sativus, L.  Garden Radish.

---

# CAPPARIDACEÆ.—The Caper Family.

POLANISIA. RAF.

P. graveolens, Raf.  Ill-scented Polanisia.

---

# VIOLACEÆ.—The Violet Family.

SOLEA, GING, DC.

S. concolor, Ging.  Green Violet.

VIOLA. L.

V. cucullata, Ait.  Common Blue Violet.
var. palmata, Gray.  Hand-Leaf Volet.

V. **sagittata**, Ait.   Arrow-leaf violet.
V. **rostrata**, Pursh.   Long-spurred violet.
V. **striata**, Ait.   Pale Violet.
V. **Canadensis**, L.   Canada Violet.
V. **pubescens**, Ait.   Yellow Violet.
var. **eriocarpa**, Nutt.   Large Yellow Violet.
V. **tricolor**, L.
var. **arvensis**, Gray, Field Violet.

---

## HYPERICACEÆ—The St. Johns-wort Family.

### HYPERICUM. L.

H. **prolificum**, L.   Shrubby St. Johns-wort.
H. **perforatum**, L.   Common St. Johns-wort.
H. **corymbosum**, Muhl.

---

## CARYOPHYLLACEÆ —The Pink Family.

### SAPONARIA. L.

S. **officinalis**, L.   Bouncing Bet.

### SILENE. L.

S. **stellata**, Ait.   Starry Campion.
S. **inflata**, Smith.   Bladder Campion.
S. **Virginica**, L.   Fire Pink.
S. **antirrhina**, L.   Catchfly.

### LYCHNIS. Tourn.

L. **Githago**, Lam.   Wheat Cockle.

### ARENARIA. L.

A, **serpyllifolia**, L.   Sandwort.

## STELLARIA. L.

S. **media**, Smith.   Common Chickweed.
S. **pubera**, Michx.   Great Chickweed.

### CERASTIUM. L.

C. **vulgatum**, L.   Mouse-ear Chickweed.
C. **viscosum**, L.   Clammy Chickweed.
C. **nutans**, Raf.   Nodding Chickweed.
C. **oblongifolium**, Torr.   Long-Leaf Chickweed.

### MOLLUGO. L.

M. **verticillata**, L.   Carpet Chickweed.

## PORTULACACEÆ — The Purslane Family.

### PORTULACA, Tourn.

P. **oleracea**, L.   Common Purslane.

### CLAYTONIA. L.

C. **Virginica**, L.   Spring Beauty.

## MALVACEÆ—The Mallow Family.

### MALVA. L.

M. **rotundifolia**, L.   Common Mallow.

### NAPÆA. Clayt.

N. **dioica**, L.   Glade Mallow.

### SIDA. L.

S. **spinosa**, L.   Spinous Sida.

### ABUTILON. Tourn.

A. **Avicennæ**, Gærtn.   Velvet Leaf.

## HIBISCUS. L.

H. militaris, Cav.   Rose Mallow.
H. Trionum, L.   Bladder Ketmia.

---

## TILIACEÆ—The Linden Family.

### TILIA. L.

T. Americana, L.   Basswood.

---

## LINACEÆ—The Flax Family.

### LINUM. L.

L. usitatissimum, L.   Common Flax.

---

## GERANIACEÆ—The Geranium Family.

### GERANIUM. L.

G. maculatum, L.   Wild Geranium.

### ERODIUM. L'Her.

E. cicutarium, L'Her.   Storksbill.

### IMPATIENS. L.

I. pallida, Nutt.   Pale Touch-me-not.
I. fulva, Nutt.   Orange Touch-me-not

### OXALIS. L.

O. violacea, L.   Wood Sorrel.
O. stricta, L.   Field Sorrel.

## RUTACEÆ—The Rue Family.

### ZANTHOXYLUM. COLDEN.

Z. **Americanum**, Mill.   Prickly Ash.

### PTELEA. L.

P. **trifoliata**, L.   Hop-tree.

---

## ANACARDIACEÆ —The Sumac Family.

### RHUS. L.

R. **typhina**, L.   Staghorn Sumac.
R. **glabra**, L.   Smooth Sumac.
R. **venenata**, DC.   Poison Sumac.
R. **Toxicodendron**, L.   Poison Ivy.
R. **aromatica**, Ait.   Fragrant Sumac.

---

## VITACEÆ—The Vine Family.

### VITIS, TOURN.

V. **Labrusca**, L.   Northern Fox Grape.
V. **æstivalis**, Michx.   Summer Grape.
V. **cordifolia**, Michx.   Frost Grape.

### AMPELOPSIS, MICHX.

A. **quinquefolia**, Michx.   American Ivy.

---

## RHAMNACEÆ—The Buckthorn Family.

### FRANGULA, TOURN.

F. **Caroliniana**, Gray.   Alder Buckthorn.

CEANOTHUS, L.

C. **Americanus**, L.   New Jersey Tea.

## CELASTRACEÆ—The Staff-tree Family.

CELASTRUS, L.

C. **scandens**, L.   Climbing Bittersweet.

EUONYMUS, Tourn.

E. **atropurpureus**, Jacq.   Waahoo.

## SAPINDACEÆ—The Maple Family.

STAPHYLEA, L.

S. **trifolia**, L.   Bladder-nut.

ÆSCULUS, L.

Æ. **Hippocastanum**, L.   Horse Chestnut.
Æ. **glabra**, Willd.   Ohio Buckeye.

ACER, Tourn.

A. **saccharinum**, Wang.   Sugar Maple.
A. **dasycarpum**, Ehrhart.   Silver Maple.
A. **rubrum**, L.   Red Maple.

NEGUNDO, Mœnch.

N. **aceroides**, Mœnch.   Ash-leaved Maple.

## POLYGALACEÆ—The Polygala Family.

POLYGALA, Tourn.

P. **Senega**, L.   Seneca Snake-root.

## LEGUMINOSÆ—The Legume Family.

### TRIFOLIUM, L,

T. pratense, L.   Red Clover.
T. stoloniferum, Muhl.   Running Clover.
T. repens, L.   White Clover.

### MELILOTUS, Tourn.

M. alba, Lam.   White Sweet Clover.

### MEDICAGO, L.

M. lupulina, L.   Black Medick.

### ROBINIA, L.

R. Pseudacacia, L.   Common Locust.
R. viscosa, Vent.   Clammy Locust.
R. hispida, L.   Bristly Locust.

### WISTARIA, Nutt.

W. frutescens, DC.   Wistaria.

### DESMODIUM, DC.

D. nudiflorum, DC.   Tick Trefoil.
D. acuminatum, DC.
D. canescens, DC.
D. cuspidatum, Torr. & Gray.
D. rigidum, DC.

### LESPEDEZA, Mich.

L. capitata, Michx.   Bush Clover.

### VICIA, Tourn.

V. Cracca, L.   Vetch.

AMPHICARP.EA, ELL.

A. monoica, Nutt.   Hog Pea-nut.

CERCIS, L.

C. Canadensis, L.   Red-bud.

CASSIA, L.

C. Marilandica, L.   Wild Senna.
C. Chamæcrista, L.   Partridge Pea.

GYMNOCLADUS, LAM.

G. Canadensis, Lam.   Coffee Tree.

GLEDITSCHIA, L.

G. triacanthos, L.   Honey Locust.

---

ROSACEÆ—The Rose Family.

PRUNUS, TOURN.

P. Americana, Mar.   Wild Plum.
P. serotina, Ehr.   Wild Black Cherry.

SPIRÆA, L.

S. lobata, Murr.   Queen of the Prairie.

POTERIUM, L.

P. Canadense.   Canadian Burnet.

AGRIMONIA, TOURN.

A. Eupatoria, L.   Common Agrimony.
A. parviflora, Ait.   Small-flowered Agrimony.

GEUM, L.

G. album, Gmelin.   White Geum.

G. **Virginianum**, L.   Virginian Geum.
G. **strictum**, Ait.   Yellow Avens.
G. **vernum**, Torr. & Gray.   Spring Geum.

### POTENTILLA, L.

P. **Norvegica**, L.   Hairy Potentilla.
P. **Canadensis**, L.
var. **simplex**, Torr. & Gray.   Cinquefoil.
P. **fruticosa**, L.   Shrubby Cinquefoil.

### FRAGARIA, Tourn.

F. **Virginiana**, Ehrh.
var. **Illinoensis**, Gray.   Wild Strawberry.

### RUBUS, Tourn.

R. **villosus**, Ait.   High Blackberry.

### ROSA, Tourn.

R. **setigera**, Michx.   Climbing Rose.
R. **Carolina**, L.   Swamp Rose.
R. **lucida**, Ehrh.   Dwarf Wild Rose.

### CRATÆGUS, L.

C. **coccinea**, L.   Scarlet-fruited Thorn.
C. **tomentosa**, L.   Pear Thorn.
var. **pyrifolia**, Gray.   Pear-leaved Thorn.
var. **mollis**, Gray.   Downy-leaved Thorn.

### PYRUS, L.

P. **coronaria**, L.   Wild Crab Apple.

### AMELANCHIER, Medic.

A. **Canadensis**, Torr. & Gray.   Service Berry.

# SAXIFRAGACEÆ—The Saxifrage Family.

### RIBES, L.
R. **Cynosbati, L.** Wild Gooseberry.
R. **floridum, L.** Wild Black Currant.

### HYDRANGEA, GRONOV.
H. **arborescens, L.** Wild Hydrangea.

### PARNASSIA, TOURN.
P. **Caroliniana, Michx.** Grass of Parnassus.

### SAXIFRAGA, L.
S. **Virginiensis, Michx.** Early Saxifrage.
S. **Pennsylvanica, L.** Swamp Saxifrage.

### HEUCHERA, L.
H. **Americana, L.** Alum-root.

### MITELLA, TOURN.
M. **diphylla, L.** Bishop's Cap.

---

# CRASSULACEÆ—The Sedum Family.

### PENTHORUM, GRONOV.
P. **sedoides, L.** Ditch Stone-crop.

### SEDUM, TOURN.
S. **acre, L.** Mossy Sedum.
S. **pulchellum, Michx.** Purple Sedum.
S. **ternatum, Michx.** White Sedum.
S. **Telephium, L.** Liver-for-ever.

# ONAGRACEÆ—The Evening Primrose Family.

### CIRCÆA, Tourn.

C. **Lutetiana**, L.   Enchanter's Nightshade.

### GAURA, L.

G. **biennis**, L.   Downy Gaura.

### ŒNOTHERA, L.

Œ. **biennis**, L.   Evening Primrose.
Œ. **fruticosa**, L.   Sun-drops.

### LUDWIGIA, L.

L. **palustris**, Ell.   Water Purslane.

---

# LYTHRACEÆ—The Lythrum Family.

### LYTHRUM, L.

L. **alatum**, Pursh.   Winged Lythrum.

---

# PASSIFLORACEÆ—The Passion-flower Family

### PASSIFLORA, L.

P. **lutea**, L.   Yellow Passion Flower.

---

# CUCURBITACEÆ—The Gourd Family.

### SICYOS, L.

S. **angulatus**, L.   Star Cucumber.

### ECHINOCYSTIS, Torr. & Gray.

E. **lobata**, Torr. & Gray.   Wild Balsam Apple.

# UMBELLIFERÆ—The Umbel Family.

### SANICULA, Tourn.

S. **Canadensis**, L. Canadian Sanicle.
S. **Marilandica**, L. Maryland Sanicle.

### PASTINACA, Tourn.

P. **sativa**, L. Wild Parsnip.

### THASPIUM, Nutt.

T. **barbinode**, Nutt. Meadow Parsnip.

### CICUTA, L.

C. **maculata**, L. Water Hemlock.

### CRYPTOTÆNIA, DC.

C. **Canadensis**, DC. Honewort.

### CHÆROPHYLLUM, L.

C. **procumbens**, Lam. Wild Chervil.

### OSMORRHIZA, Raf.

O. **longistylis**, DC. Smooth Sweet Cicely.
O. **brevistylis**, DC. Hairy Sweet Cicely.

### CONIUM, L.

C. **maculatum**, L. Poison Hemlock.

### ERIGENIA, Nutt.

E. **bulbosa**, Nutt. Harbinger of Spring.

# ARALIACEÆ—The Ginseng Family.

### ARALIA, Tourn.

A. **racemosa**, L. Spikenard.
A. **quinquefolia**, Gray. Ginseng.

## CORNACEÆ—The Dogwood Family.

### CORNUS, TOURN.

C. **florida,** L.    Flowering Dogwood.
C. **stolonifera,** Michx.    Red Osier Dogwood.
C. **paniculata,** L'Her.    Bush Dogwood.

### NYSSA, L.

N. **multiflora,** Wang.    Black Gum Tree.

*DIVISION II.*—MONOPETALOUS.

## CAPRIFOLIACEÆ—The Honeysuckle Family

### LONICERA, L.

L. **sempervirens,** Ait.    Trumpet Honeysuckle.

### TRIOSTEUM, L.

T. **perfoliatum,** L.    Fever-wort.

### SAMBUCUS, TOURN.

S. **Canadensis,** L.    Black Elder.

### VIBURNUM, L.

V. **Lentago,** L.    Sweet Viburnum.
V. **prunifolium,** L.    Black Haw.
V. **pubescens,** Pursh.    Downy Viburnum.

## RUBIACEÆ—The Madder Family.

### GALIUM, L.

G. **Aparine,** L.    Cleavers.

G. concinnum, Torr. & Gray.
G. triflorum, Michx. Sweet-scented Galium.
G. circæzans, Michx. Wild Liquorice.

CEPHALANTHUS, L.

C. occidentalis, L. Button-bush.

MITCHELLA, L.

M. repens, L. Partridge Berry.

HOUSTONIA, L.

H. purpurea, L. Purple Houstonia.

----

VALERIANACEÆ—The Valerian Family.

VALERIANA, TOURN.

V. pauciflora, Michx. Wild Valerian.

FEDIA, GÆRTN.

F. Fagopyrum, Torr. & Gray. Lamb Lettuce.

----

DIPSACEÆ—The Teasel Family.

DIPSACUS, TOURN.

D. sylvestris, Mill. Wild Teasel.

----

COMPOSITÆ—The Composite Family.

VERNONIA, SCHREB.

V. fasciculata, Michx. Ironweed.

ELEPHANTOPUS, L.

E. Carolinianus, Willd. Elephant's-foot.

## EUPATORIUM, Tourn.

E. **purpureum**, L.   Joe Pye Weed.
E. **perfoliatum**, L.   Thoroughwort.
E. **ageratoides**, L.   White Snake-root.

## CONOCLINIUM, DC.

C. **cœlestinum**, DC.   Mist-flower.

## ASTER, L.   Aster.

A. **azureus**, Lindl.     A. **cordifolius**, L.
A. **Tradescanti**, L.     A. **miser**, L.
A. **puniceus**, L.

## ERIGERON, L.

E. **Canadense**, L.   Horse-weed.
E. **bellidifolium**, Muhl.   Robin's Plantain.
E. **Philadelphicum**, L.   Common Fleabane.
E. **annuum**, Pers.   White-weed.

## SOLIDAGO, L.   Golden-rod.

S. **latifolia**, L.     S. **cæsia**, L.
S. **Ohioensis**, Rid.     S. **Riddellii**, Frank.
S. **patula**, Muhl.     S. **Canadensis**, L.
S. **gigantea**, Ait.

## INULA, L.

I. **Helenium**, L.   Elecampane.

## POLYMNIA, L.

P. **Canadensis**, L.   Leaf Cup.

## SILPHIUM, L.

S. **terebinthinaceum**, L.   Prairie Dock.
S. **trifoliatum**, L.

## AMBROSIA, Tourn.

A. **trifida**, L. Great Rag-weed.
A. **artemisiæfolia**, L. Rag-weed.

## XANTHIUM, Tourn.

X. **strumarium**, L. Cocklebur.

## ECLIPTA, L.

E. **procumbens**, Michx.

## HELIOPSIS, Pers.

H. **lævis**, Pers. Oy-eye.

## ECHINACEA, Mœnch.

E. **purpurea**, Mœnch. Purple Cone-flower.

## RUDBECKIA, L.

R. **laciniata**, L.      R. **triloba**, L.
R. **hirta**, L. Cone-flower.

## LEPACHYS, Raf.

L. **pinnata**, Torr. & Gray.

## HELIANTHUS, L.

H. **annuus**, L. Garden Sun-flower.
H. **microcephalus**, Torr. & Gray.
H. **giganteus**, L.
H. **strumosus**, L.
II. **hirsutus**, Raf.
H. **decapetalus**, L.
H. **doronicoides**, Lam.

## ACTINOMERIS, Nutt.

A. **squarrosa**, Nutt.
A. **helianthoides**, Nutt.

COREOPSIS, L.

C. aristosa, Michx.

BIDENS, L.

B. frondosa, L.   Spanish Needles.
B. chrysanthemoides, Michx.
B. bipinnata, L.

HELENIUM, L.

H. autumnale, L.   Sneeze-weed.

MARUTA, Cass.

M. Cotula, DC.   Dog Fennel.

ACHILLEA, L.

A. Millefolium, L.   Yarrow.

LEUCANTHEMUM, Tourn.

L. vulgare, Lam.   Ox-eye Daisy.

TANACETUM, L.

T. vulgare, L.   Tansy.

ANTENNARIA, Gærtn.

A. plantaginifolia, Hook.   Everlasting.

ERECHTHITES, Raf.

E. hieracifolia, Raf.   Fireweed.

CACALIA, L.

C. atriplicifolia, L.   Indian Plantain.
C. tuberosa, Nutt.

SENECIO, L.

S. aureus, L.   Golden Senecio.

CENTAUREA, L.

C. **Cyanus**, L. Bluebottle.

CIRSIUM, Tourn.

C. **lanceolatum**, Scop. Common Thistle.

C. **discolor**, Spreng.

C. **altissimum**, Spreng.

C. **muticum**, Michx. Swamp Thistle.

ONOPORDON, Vaill.

O. **acanthium**, L. Scotch Thistle.

LAPPA, Tourn.

L. **officinalis**, All. Burdock.

HIERACIUM, Tourn.

H. **venosum**, L. Rattlesnake Weed.

NABALUS, Cass.

N. **albus**, Hook.

N. **racemosus**, Hook.

TARAXACUM, Haller.

T. **Dens-leonis**, Desf. Dandelion.

LACTUCA, Tourn.

L. **Canadensis**, L. Wild Lettuce.

MULGEDIUM, Cass.

M. **acuminatum**, DC. Blue Lettuce.

M. **Floridanum**, DC.

M. **leucophæum**, DC.

SONCHUS, L.

S. **oleraceus**, L. Common Sow Thistle.

S. **asper**, Vill. Spiny-leaved Sow Thistle.

# LOBELIACEÆ—The Lobelia Family.

### LOBELIA, L.

L. cardinalis, L.   Cardinal Flower.
L. syphilitica, L.   Great Lobelia.
L. inflata, L.   Indian Tobacco.

---

# CAMPANULACEÆ—The Campanula Family.

### CAMPANULA, TOURN.

C. aparinoides, Pursh.   Marsh Bellflower.
C. Americana, L.   Tall Bellflower.

### SPECULARIA, HEISTER.

S. perfoliata, A. DC.   Venus's Looking-glass.

---

# ERICACEÆ—The Heath Family.

### MONOTROPA, L.

M. uniflora, L.   Indian Pipe.

---

# EBENACEÆ—The Ebony Family.

### DIOSPYROS, L.

D. Virginiana, L.   Persimmon.

---

# PLANTAGINACEÆ — The Plantain Family.

### PLANTAGO, L.

P. major, L.   Common Plantain.
P. lanceolata, L.   English Plantain.

# RRIMULACEÆ—The Primrose Family.

### DODECATHEON, L.

D. **Meadia**, L. Shooting Star.

### LYSIMACHIA, Tourn.

L. **ciliata**, L. Loosestrife.
L. **longifolia**, Pursh.

### SAMOLUS, L.

S. **Valerandi**, L.
var. **Americanus**, Gray. Water Pimpernel.

---

# BIGNONIACEÆ—The Bignonia Family.

### TECOMA, Juss.

T. **radicans**, Juss. Trumpet Creeper.

### CATALPA, Scop.

C. **bignonioides**, Walt. Catalpa Tree.

### MARTYNIA, L.

M. **proboscidea**, Glox. Devil's Claws.

---

# OROBANCHACEÆ—The Beech-drops Family.

### EPIPHEGUS, Nutt.

E. **Virginiana**, Bart. Beech-drops.

### CONOPHOLIS, Wall.

C. **Americana**, Wall. Cancer-root.

# SCROPHULARIACEÆ—The Figwort Family.

### VERBASCUM, L.

V. **Thapsus**, L.    Common Mullein.
V. **Blattaria**, L.    Moth Mullein.

### LINARIA, Tourn.

L. **vulgaris**, Mill.    Butter-and-eggs.

### SCROPHULARIA, Tourn.

S. **nodosa**, L.    Figwort.

### COLLINSIA, Nutt.

C. **verna**, Nutt.    Blue-eyed Mary.

### CHELONE, Tourn.

C. **glabra**, L.    Turtle-head.

### PENTSTEMON, Mitchell.

P. **pubescens**, Sol.    Hairy Pentstemon.
P. **Digitalis**, Nutt.    Smooth Pentstemon.

### MIMULUS, L.

M. **ringens**, L.    Monkey-flower.
M. **alatus**, Ait.

### CONOBEA, Aublet,

C. **multifida**, Benth.

### ILYSANTHES, Raf.

I. **gratioloides**, Benth.    False Pimpernel.

### VERONICA, L.

V. **Virginica**, L.    Culver's-root.
V. **Anagallis**, L.    Water Speedwell.

V. officinalis, L.   Common Speedwell.
V. serpyllifolia, L.   Thyme-leaved Speedwell.
V. peregrina, L.   Purslane Speedwell.
V. arvensis, L.   Corn Speedwell.

### SEYMERIA, Pursh.

S. macrophylla, Nutt.   Mullein Foxglove.

### GERARDIA, L.

G. purpurea, L.   Purple Gerardia.
G. tenuifolia, Vahl.   Slender Gerardia.

### PEDICULARIS, Tourn.

P. lanceolata, Michx.   Lousewort.

----

## ACANTHACEÆ—The Acanthus Family.

### DIANTHERA, Gronov.

D. Americana, L.   Water Willow.

### RUELLIA, L.

R. ciliosa, Pursh.         R. strepens, L.

----

## VERBENACEÆ—The Verbena Family.

### VERBENA, L.

V. hastata, L.   Blue Vervain.
V. urticifolia, L.   White Vervain.
V. stricta, Vent.   Hoary Vervain.
V. bracteosa, Michx.
V. Aubletia, L.

## LIPPIA, L.

L. lanceolata, Michx,   Fog-fruit.

## PHRYMA, L.

P. Leptostachya, L.   Lopseed.

---

## LABIATÆ—The Mint Family.

### TEUCRIUM, L.

T. Canadense, L.   American Germander.

### ISANTHUS, Michx.

I. cæruleus, Michx.   False Pennyroyal.

### MENTHA, L.

M. viridis, L.   Spearmint.
M. piperita, L.   Peppermint.
M. Canadensis, L.   Wild Mint.

### LYCOPUS, L.

L. Europæus, L.   Water Hoarhound.

### PYCNANTHEMUM, Michx.

P. muticum, Pers.   Mountain Mint.
P. lanceolatum, Pursh.

### MELISSA, L.

M. officinalis, L.   Common Balm.

### HEDEOMA, Pers.

H. pulegioides, Pers.   Pennyroyal.

### MONARDA, L.

M. didyma, L.   Oswego Tea.

M. fistulosa, L. Wild Bergamot.
M. Bradburiana, Beck.

LOPHANTHUS, Benth.

L. scrophulariæfolius, Benth. Giant Hyssop.

NEPETA, L.

N. Cataria, L. Catnip.
N. Glechoma, Benth. Ground Ivy.

SYNANDRA, Nutt.

S. grandiflora, Nutt.

PHYSOSTEGIA, Benth.

P. Virginiana, Benth. Dragon-head.

BRUNELLA, Tourn.

B. vulgaris, L. Common Self-heal.

SCUTELLARIA, L. Skull Cap.

S. versicolor, Nutt.     S. canescens, Nutt.
S. parvula, Michx.     S. lateriflora, L.

MARRUBIUM, L.

M. vulgare, L. Common Hoar-hound.

GALEOPSIS, L.

G. Tetrahit, L. Hemp Nettle.

STACHYS, L.

S. palustris, L. Hedge Nettle.

LEONURUS, L.

L. Cardiaca, L. Motherwort.

LAMIUM, L.
L. amplexicaule, L.    Dead Nettle.

---

## BORRAGINACEÆ—The Borage Family.

ECHIUM, Tourn.
E. vulgare, L.    Blue-weed.

ONOSMODIUM, Michx.
O. Carolinianum, DC.    False Gromwell.

LITHOSPERMUM, Tourn.
L. arvense, L.    Field Gromwell.
L. latifolium, Michx.    Wild Gromwell.

MERTENSIA, Roth.
M. Virginica, DC.    Lungwort.

MYOSOTIS, L.
M. arvensis, Hoffm.    Forget-me-not.

ECHINOSPERMUM, Swartz.
E. Lappula, Lehm.    Stick-seed.

CYNOGLOSSUM, Tourn.
C. officinale, L.    Hound's Tongue.
C. Virginicum, L.    Wild Comfrey.
C. Morisoni, DC.    Beggar's Lice.

---

## HYDROPHYLLACEÆ—The Water-leaf Family

HYDROPHYLLUM, L.
H. macrophyllum, Nutt.    Large-leaved H.

H. **Virginicum,** L.. Virginian Hydrophyllum.
H. **appendiculatum,** Michx.

PHACELIA, Juss.

P. **Purshii,** Buckley. Miami Mist.

---

# POLEMONIACEÆ—The Phlox Family.

POLEMONIUM, Tourn.

P. **reptans,** L. Jacob's Ladder.

PHLOX, L.

P. **paniculata,** L.. Bunch Phlox.
P. **maculata,** L. Spotted-stem Phlox.
P. **glaberrima,** L.. Smooth Phlox.
P. **divaricata,** L.. Sweet William.
P. **subulata,** L. Moss Pink.

---

# CONVOLVULACEÆ—The Morning-glory Family.

QUAMOCLIT, Tourn,

Q. **vulgaris,** Choisy. Cypress Vine.

IPOMŒA, L.

I. **purpurea,** Lam. Purple Morning-glory.
I. **Nil,** Roth. Blue Morning-glory.

CALYSTEGIA, R. Br.

C. **sepium,** R. Br. Hedge Bindweed.

## SOLANACEÆ—The Potato Family.

### SOLANUM, Tourn.

S. **Dulcamara**, L. Bittersweet.
S. **nigrum**, L. Nightshade.
S. **Carolinense**, L. Horse Nettle.

### PHYSALIS, L.

P. **viscosa**, L. Viscid Ground Cherry.
P. **Pennsylvanica**, L. Smooth Ground Cherry.

### NICANDRA, Adans.

N. **physaloides**, Gærtn. Apple of Peru.

### LYCIUM, L.

L. **vulgare**, Dunal. Matrimony Vine.

### HYOSCYAMUS, Tourn.

H. **niger**, L. Black Henbane.

### DATURA, L.

D. **Stramonium**, L. Green-stem Jimson.
D. **Tatula**, L. Purple-stem Jimson.

## GENTIANACEÆ—The Gentian Family.

### SABBATIA, Adans.

S. **angularis**, Pursh. American Centaury.

### FRASERA, Walt.

F. **Carolinensis**, Walt. American Columbo.

### GENTIANA, L.

G. **crinita**, Frœl. Fringed Gentian.

# LOGANIACEÆ—The Logania Family.

## SPIGELIA, L.

S. Marilandica, L.  Pink-root.

----

# APOCYNACEÆ—The Dogbane Family.

## APOCYNUM, Tourn.

A. androsæmifolium, L.  Dogbane.
A. cannabinum, L.  Indian Hemp.

----

# ASCLEPIADACEÆ—The Milk-weed Family.

## ASCLEPIAS, L.

A. Cornuti, Decaisne.  Common Milk-weed.
A. phytolaccoides, Pursh.  Poke Milk-weed.
A. purpurascens, L.  Purple Milk-weed.
A. quadrifolia, Jacq.  Four-leaved Milk-weed.
A. incarnata, L.  Swamp Milk-weed.
A. tuberosa, L.  Pleurisy-root.

## ENSLENIA, Nutt.

E. albida, Nutt.

## GONOLOBUS, Michx.

G. lævis, Michx.

----

# OLEACEÆ—The Olive Family.

## LIGUSTRUM, Tourn.

L. vulgare, L.  Privet.

### CHIONANTHUS, L.
C. **Virginica,** L.  Fringe-tree.

### FRAXINUS, Tourn.
F. **Americana,** L.  White Ash.
F. **viridis,** Michx.  Green Ash.
F. **sambucifolia,** Lam.  Black Ash.
F. **quadrangulata,** Michx.  Blue Ash.

## DIVISION III.—APETALOUS.

## ARISTOLOCHIACEÆ—The Wild Ginger Family.

### ASARUM, Tourn.
A. **Canadense,** L.  Wild Ginger.

## PHYTOLACCACEÆ—The Pokeweed Family.

### PHYTOLACCA, Tourn.
P. **decandra,** L.  Pokeweed.

## CHENOPODIACEÆ—The Pigweed Family.

### CHENOPODIUM, L.
C. **album,** L.  Common Pigweed.
C. **hybridum,** L.  Maple-leaved Pigweed.
C. **Botrys,** L.  Jerusalem Oak.

# AMARANTACEÆ—The Amaranth Family.

AMARANTUS, Tourn.

A. retroflexus, L. Green Amaranth.
A. albus, L. White Amaranth.
A. spinosus, L. Thorny Amaranth.

---

# POLYGONACEÆ—The Polygonum Family.

POLYGONUM, L-

P. incarnatum, Ell.
P. Persicaria, L. Lady's Thumb.
P. Hydropiper, L. Smartweed.
P. Virginianum, L.
P. aviculare, L. Knotgrass.
var. erectum, Roth. Erect Knotgrass.
P. dumetorum, L. Climbing Buckwheat.

FAGOPYRUM, Tourn.

F. esculentum, Mœnch. Buckwheat.

RUMEX, L.

R. crispus, L. Curled Dock.
R. obtusifolius, L. Bitter Dock.
R. Acetosella, L. Sheep Sorrel.

---

# LAURACEÆ—The Laurel Family.

SASSAFRAS, Nees.

S. officinale, Nees. Sassafras.

LINDERA, Thunberg.

L. Benzoin, Meisner. Spice-bush.

## THYMELEACEÆ—The Mezereum Family.

### DIRCA L.

D. palustris, L.   Leatherwood.

---

## LORANTHACEÆ—The Mistletoe Family.

### PHORADENDRON, Nutt.

P. flavescens, Nutt.   American Mistletoe.

---

## SAURURACEÆ—The Lizard's Tail Family.

### SAURURUS. L.

S. cernuus, L.   Lizard's Tail.

---

## EUPHORBIACEÆ—The Euphorbia Family.

### EUPHORBIA, L.

E. humistrata, Eng.   E. hypericifolia, L.
E. obtusata, Pursh.   E. commutata, Eng.

### ACALYPHA, L.

A. Virginica, L.   Three-seeded Mercury.

---

## URTICACEÆ—The Nettle Family.

### ULMUS, L.

U. fulva, Mich.   Slippery Elm.
U. Americana, L.   American Elm.

### CELTIS, Tourn.

C. occidentalis, L.   Hackberry.

## MORUS, Tourn.

M. rubra, L. Mulberry.

## URTICA, Tourn.

U. gracilis, Ait. Tall Nettle.

## LAPORTEA, Gaudichaud.

L. Canadensis, Gaudichaud. Wood Nettle.

## PILEA, Lindl.

P. pumila, Gray. Clearweed.

## PARIETARIA, Tourn.

P. Pennsylvanica, Muhl. Pellitory.

## CANNABIS, Tourn.

C. sativa, L. Hemp.

---

# PLATANACEÆ—The Sycamore Family.

## PLATANUS, L.

P. occidentalis, L. American Sycamore.

---

# JUGLANDACEÆ—The Walnut Family.

## JUGLANS, L.

J. cinerea, L. Butternut.
J. nigra, L. Black Walnut.

## CARYA, Nutt.

C. alba, Nutt. Shell-bark Hickory.
C. sulcata, Nutt.
C. tomentosa, Nutt.
C. amara, Nutt. Bitter-nut Hickory.

## CUPULIFERÆ—The Oak Family.

### QERCUS, L.

Q. alba, L.    White Oak.
Q. macrocarpa, Michx.    Bur Oak.
Q. Prinus, L.    Chestnut Oak.
Q. imbricaria, Michx.    Laurel Oak.
Q. coccinea, Wang.    Scarlet Oak.
Q. rubra, L.    Red Oak.

### FAGUS, Tourn.

F. ferruginea, Ait.    American Beech.

### CORYLUS, Tourn.

C. Americana, Walt.    Hazel Nut.

### OSTRYA, Mich.

O. Virginica, Willd.    Iron-wood.

### CARPINUS, L.

C. Americana, Michx.    Water Beech.

---

## SALICACEÆ—The Willow Family.

### SALIX, Tourn.

S. discolor, Muhl.    Glaucous Willow.
S. alba, L.    White Willow.
S. Babylonica, Tourn.    Weeping Willow.

### POPULUS, Tourn.

P. monilifera, Ait.    Cotton-wood.
P. balsamifera, L.
var. candicans, Gray.    Balm of Gilead.

P. **dilatata, Ait.** Lombardy Poplar.
P. **alba, L.** Silver Poplar.

*Sub-class II.*—GYMNOSPERMÆ.

**CONIFERÆ**—The Pine Family.

THUJA, Tourn.
T. **occidentalis, L.** Arbor Vitæ.

JUNIPERUS, L.
J. **Virginiana, L.** Red Cedar.

*CLASS II.*

MONOCOTYLEDONOUS OR ENDOGENOUS PLANTS.

**ARACEÆ**—The Arum Family.

ARISÆMA, Martius.
A. **triphyllum, Torr.** Indian Turnip.
A. **Dracontium, Schott.** Dragon-root.

SYMPLOCARPUS, Salisb.
S. **fœtidus, Salisb.** Skunk Cabbage.

ACORUS, L.
A. **Calamus, L.** Sweet Flag.

## LEMNACEÆ—The Duck-weed Family.

### LEMNA, L.

L. perpusilla, Torr.   Duck-weed.

----

## TYPHACEÆ—The Cat-tail Family.

### TYPHA, Tourn.

T. latifolia, L.   Cat-tail.

### SPARGANIUM, Tourn.

S. eurycarpum, Eng.   Bur-reed.

----

## NAIADACEÆ—The Pond-weed Famiiy.

### POTAMOGETON, Tourn.

P. natans, L.   Pond-weed.

----

## ALISMACEÆ—The Water Plantain Family.

### ALISMA, L.

A. Plantago, L.
var. Americanum.   Water Plantain.

### SAGITTARIA, L.

S. variabilis, Eng.   Arrow-head.

----

## ORCHIDACEÆ—The Orchis Family.

### ORCHIS, L.

O. spectabilis, L.   Showy Orchis.

### HABENARIA, Willd.

H. **viridis**, R. Br.
var. **bracteata**, Reich. Green Orchis.
H. **leucophæa**, Nutt. White-fringed Orchis,
H. **psycodes**, Gray. Purple-fringed Orchis,

### SPIRANTHES, Richard.

S. **latifolia**, Torr. Ladies' Tresses.

### POGONIA, Juss.

P. **pendula**, Lindl.

### LIPARIS, Richard.

L. **Lœselii**, Richard. Twayblade.

### CORALLORHIZA, Haller.

C. **ondontorhiza**, Nutt. Coral-root.

### APLECTRUM, Nutt.

A. **hyemale**, Nutt. Putty-root.

### CYPRIPEDIUM, L.

C. **pubescens**, Willd. Yellow Lady's Slipper.
C. **spectabile**, Swartz. Showy Lady's Slipper.

---

## AMARYLLIDACEÆ—The Amaryllis Family.

### HYPOXYS, L.

H. **erecta**, L. Star-grass.

---

## IRIDACEÆ—The Iris Family.

### IRIS, L.

I. **versicolor**, L. Large Blue Flag.

SISYRINCHIUM, L.

S. Bermudiana, L.   Blue-eyed Grass.

---

SMILACEÆ—The Smilax Family.

SMILAX, TOURN.

S. rotundifolia.   Greenbrier.
S. hispida, Muhl.
S. herbacea, L.   Carrion Flower.

---

LILIACEÆ—The Lily Family.

TRILLIUM, L.

T. sessile, L.   Purple Trillium.
T. grandiflorum, Salisb.   Large White Trillium.
T. cernuum, L.   Nodding Trillium.

UVULARIA, L.

U. grandiflora, Smith.   Bellwort.

SMILACINA, DESF.

S. racemosa, Desf.   False Solomon's Seal.

POLYGONATUM, TOURN.

P. biflorum, Ell.   Small Solomon's Seal.
P. giganteum, Diet.   Great Solomon's Seal.

ASPARAGUS, L.

A. officinalis, L.

LILIUM, L.

L. Canadense, L.   Wild Yellow Lily.

ERYTHRONIUM, L.

E. **Americanum**, Smith.   Yellow Adder Tongue.
E. **albidum**, Nutt.   White Adder Tongue.

SCILLA, L.

S. **Fraseri**, Gray.   Wild Hyacinth.

ALLIUM, L.

A. **cernuum**, Roth.   Wild Onion.
A. **Canadense**, Kalm.   Wild Garlic.

HEMEROCALLIS, L.

H. **fulva**, L.   Orange Day Lily.

YUCCA, L.

Y. **filamentosa**. L.   Adam's Needle.

## JUNCACEÆ —The Rush Family.

LUZULA, DC.   Wood Rush.

L. **campestris**, DC.

JUNCUS, L.   Bog Rush.

J. **bufonius**, L.       J. **tenuis**, Willd.

## COMMELYNACEÆ—The Spiderwort Family.

TRADESCANTIA, L.   Spiderwort.

T. **Virginica**, L.       T. **pilosa**, Lehm.

# CYPERACEÆ—The Sedge Family.

## CYPERUS, L.

C. inflexus, Muhl.          C. phymatodes, Muhl.

## ELEOCHARIS, R. Br.

E. obtusa, Sch.          E. palustris, R. Br.
E. tenuis, Sch.

## SCIRPUS, L.

S. validus, Vahl.

## CAREX, L.

C. polytrichoides, Muhl.  C. grisea, Wahl.
C. Steudelii, Kunth.      C. gracillima, Schw.
C. bromoides, Schk.       C. triceps, Michx.
C. siccata, Dew.          C. plantaginea, Lam.
C. teretiuscula, Good.    C. laxiflora, Lam.
C. decomposita, Muhl.     var. latifolia, Boot.
C. vulpinoidea, Michx.    var. blanda, Gray.
C. stipata, Muhl.         C. Hitchcockiana.
C. sparganioides, Muhl.   C. Pennsylvanica, Lam.
C. cephalophora, Muhl.    C. varia, Muhl.
C. rosea, Schk.           C. pubescens, Muhl.
C. lagopodioides, Schk.   C. flava, L.
C. fœnea, Willd. ?        C. filiformis, L.
var. sabulonum,  ?        C. riparia, Curtis.
C. straminea, Schk.       C. trichocarpa, Muhl.
C. stricta, Lam.          C. hystricina, Willd.
C. Shortiana, Dew.        C. tentaculata, Muhl.
C. granularis, Muhl.      C. stenolepis, Torr.

# GRAMINEÆ—The Grass Family.

### LEERSIA, SOLANDER,

L. **oryzoides,** Swartz.

### PHLEUM, L.

P. **pratense,** L. Timothy.

### AGROSTIS, L.

A. **vulgaris,** With. Herd's Grass.

### ELEUSINE, GÆRTN.

E. **Indica,** Gærtn. Crab Grass.

### TRICUSPIS, BEAUV.

T. **seslerioides,** Torr. Tall Red-top.

### POA, L.

P. **pratensis,** L. Kentucky Blue Grass.

### BROMUS, L.

B. **secalinus,** L. Cheat.

### ARUNDINARIA, MICHX.

A. **tecta,** Muhl. Small Cane.

### PANICUM, L.

P. **sanguinale,** L. Finger Grass.
P. **Crus-galli,** L. Barn-yard Grass.

### SETARIA, BEAUV.

S. **viridis,** Beauv. Green Foxtail.

## SERIES II.

## CRYPTOGAMOUS OR FLOWERLESS PLANTS.

---

### EQUISETACEÆ—The Horsetail Family.

EQUISETUM, L.

E. arvense, L. Common Horsetail.
E. hyemale, L. Scouring Rush.

---

### FILICES—The Ferns.

POLYPODIUM, L.

P. vulgare, L.            P. incanum, Swartz.

ADIANTUM, L.

A. pedatum, L.

PELLÆA, LINK.

P. atropurpurea, Link.

ASPLENIUM, L.

A. Trichomanes, L.
A. ebeneum, Ait.
A. Ruta-muraria, L.
A. angustifolium, Mich.
A. thelypteroides, Michx.
A. Filix-fœmina, Bernh.

CAMPTOSORUS. LINK.

C. rhizophyllus, Link. Walking Fern.

PHEGOPTERIS, Fee.

P. hexagonoptera, Fee.

ASPIDIUM, Swartz.

A. Thelypteris, Swartz.
A. spinulosum, Swartz.
var. intermedium, D. C. Eat.
A. cristatum, Swartz.
A. Goldianum, Hook.
A. marginale, Swartz.
A. acrostichoides, Swartz.
var. incisum, Gray.

CYSTOPTERIS, Bernh.

C. bulbifera, Bernh.
C. fragilis, Bernh.
var. dentata, Hook.
var. angustata.

ONOCLEA, L.

O. sensibilis, L.

WOODSIA, R. Brown,

W. obtusa, Torr.

OSMUNDA, L.

O. cinnamomea, L.

BOTRYCHIUM, Swartz.

B. Virginicum, Swartz.
B. ternatum, Swartz.
var. lunarioides, Swartz.
var. dissectum, Milde.

OPHIOGLOSSUM, L.

O. **vulgatum**, L. Adder-tongue Fern.

## LYCOPODIACEÆ—The Club-mosses.

LYCOPODIUM, L.

L. **lucidulum**, Michx.

## MUSCI—The Mosses.

SPHAGNUM, DILL.

S. **cymbifolium**, Dill.    S. **acutifolium**, Ehrh.

GYMNOSTOMUM, HEDW.

G. **curvirostrum**, Hedw.

WEISIA, HEDW.

W. **viridula**, Brid.

DICRANUM, HEDW.

D. **varium**, Hedw.           D. **montanum**, Hedw.
D. **rufescens**, Turn.        D. **flagellare**, Hedw.
D. **heteromallum**,          D. **scoparium**, L.
var. **orthocarpon.**         var. **orthophyllum,**

CERATODON, BRID.

C. **purpureum**, Brid.

LEUCOBRYUM, HAMPE.

L. **glaucum**, Hampe.

FISSIDENS, HEDW.

F. **osmundioides**, Hedw.  F. **taxifolius**, Hedw.
F. **subbasilaris**, Hedw.   F. **adiantoides**, Hedw.

TRICHOSTOMUM, Br. & Sch.

T. pallidum, Hedw.

BARBULA, Hedw.

B. unguiculata, Hedw.   B. cæspitosa, Schwægr.

DIDYMODON, Br. & Sch.

D. rubellus, Br. & Sch.

POTTIA, Ehrh.

P. truncata, Br. & Sch.

TETRAPHIS, Hedw.

T. pellucida, Hedw.

DRUMMONDIA, Hook.

D. clavellata, Hook.

ORTHOTRICHUM, Hedw.

O. strangulatum, Beauv.

GRIMMIA, Ehrh.

G. Pennsylvanica, Schwægr.

HEDWIGIA, Ehrh.

H. ciliata, Ehrh.

ATRICHUM, Beauv.

A. angustatum, Beauv.

POLYTRICHUM, Brid.

P. formosum, Hedw.

TIMMIA, Hedw.

T. megapolitana. Hedw.

AULACOMNION, Schwægr.

A. heterostichum, Br. & Sch.

BRYUM, Br. & Sch.

B. pyriforme, Hedw.　　B. bimum, Schreb.
B. roseum, Schreb.　　B. intermedium, Brid.
B. albicans, Wahlenb.　B. uliginosum, Br.
B. argenteum, Linn.
B. pseudo-triquetrum, Schwægr.

MNIUM, Br. & Sch.

M. affine, Bland.　　M. cuspidatum, Hedw.

BARTRAMIA, Hedw.

B. pomiformis, Hedw.

FUNARIA, Schreb.

F. hygrometrica, Hedw.

PHYSCOMITRIUM, Brid.

P. pyriforme, Br. & Sch.

APHANORHEGMA, Sulliv.

A. serrata, Sulliv.

LEUCODON, Schwægr.

L. julaceus, Sulliv.

LEPTODON, Mohr.

L. trichomitrion, Mohr.

ANOMODON, Hook & Tayl.

A. obtusifolius, Br. & Sch.
A. attenuatus, Hub.
A. tristis, Cesati.

LESKEA, Hedw.; Bry. Eu.

L. rostrata, Hedw.    L. denticulata, Sulliv.

THELIA, Sulliv.

T. hirtella, Sulliv.

MYURELLA, Bry. Eu.

M. Careyana, Sulliv.

THUIDIUM.

T. minutulum.    T. æstivum.

PYLAISÆA, Bry. Eu.

P. intricata, Bry. Eu.

PLATYGYRIUM, Bry. Eu.

P. repens, Bry. Eu.

CYLINDROTHECIUM, Bry. Eu.

C. cladorrhizans, Bry. Eu.
C. seductrix, Bry. Eu.
C. brevisetum, Bry. Eu.

NECKERA, Hedw.; Bry. Eu.

N. pennata, Hedw.

CLIMACIUM, Web. & Mohr.

C. Americanum, Brid.

HYPNUM, Dill.

H. tamariscinum,        H. filicinum, L.
H. delicatulum, L.       H. molluscum, Hedw.
H. minutulum, Hedw.    H. imponens, Hedw.
H. gracile, Br. & Sch.    H. curvifolium, Hedw.
H. triquetrum, L.         H. rugosum, Ehrh.

H. Alleghaniense,    H. lætum, Brid.
H. hians, Hedw.    H. acuminatum, Beauv.
H. Sullivantii, Spruce,    H. rutabulum, L.
H. Boscii, Schwægr.    H. hispidulum, Brid.
H. serrulatum, Hedw.    H. adnatum, Hedw.
H. deplanatum,    H. radicale, Brid.
H. demissum, Wils.    H. orthocladon, Beauv.
H. microcarpum,    H. riparium, Hedw.
H. cylindricarpum,    H. confervoides, Schu.
H. Schreberi, Willd.    II. chrysophyllum.
H. aduncum, Hedw.

## HEPATICÆ—The Liverworts.

### RICCIA, Mich.

R. Sullivantii, Austin.

### MARCHANTIA, L.

M. polymorpha, L.

### FEGATELLA, Raddi.

F. conica, Corda.

### ANEURA, Dumort.

A. pinguis, Dumort.    A. latifrons.
A. palmata, Nees.

### GEOCALYX, Nees.

G. graveolens, Nees.

### CHILOSCYPHUS, Corda.

C. polyanthos, Corda,
C. ascendens, Hook & Wils.

## JUNGERMANNIA, L.

J. connivens, Dick.  J. incisa, Schrader.
J. curvifolia, Dick.  J. scutata, Weber.
J. divaricata, E. Bot.  J. Schraderi, Mart.
J. bicuspidata, L.  J. crenuliformis.

## SCAPANIA, Lindenberg.

S. nemorosa, Nees.

## PLAGIOCHILA, Nees. & Montague.

P. asplenioides, Nees. & Montague.
P. porelloides, Lindenberg.

## FRULLANIA, Raddi.

F. Virginica, Lehm.  F. æolotis, Nees.
F. Eboracensis, Lehm.  F. squarrosa, Nees.

## LEJEUNIA, Libert.

L. clypeata, Schw.  L. echinata, Tayl.

## MADOTHECA, Dumor.

M. Thuja.  M. rivularis, Nees.

## RADULA, Nees.

R. obconica, Sulliv.

## PTILIDIUM, Nees.

P. ciliare, Nees.

## TRICHOCOLEA, Nees.

T. tomentella, Nees.  T. Biddlecomæa, Aust.

MASTIGOBRYUM, Nees.

M. trilobatum, Nees.

## LICHENES—Lichens.

RAMALINA, Ach,

R. calicaris, Fr.

CETRARIA, Ach.

C. ciliaris, Ach.

USNEA, Ach.

U. barbata, Fr.          U. angulata, Ach.
var. florida, Fr.
var. hirta, Fr.

THELOSCHISTES, Norm.

T. parietinus, Norm.
var. polycarpus, Fr.

·PARMELIA, DeNot.

P. perforata, Ach.       P. Borreri, Turn.
var. crinita, Tuckerm.   var. rudecta, Tuckerm.
                         P. caperata, Ach.

PHYSCIA, Fries.

P. speciosa, Fr.         P. stellaris, Nyl.
var. galactophylla.      P. detonsa, Fr.

STICTA, Delis.

S. glomerulifera, Delis.  S. pulmonaria, Ach.

## CLADONIA, Hoffm.

C. pyxidata, Fr.  
C. fimbriata, Fr.  
var. adspersa, Tuckerm.  
C. squamosa, Hoffm.  
var. delicata, Fr.

C. macilenta, Hoffm.  
C. furcata, Fr.  
C. mitrula.

## ENDOCARPON, Hedw.

E. miniatum, Schær.

# FUNGI.

## AGARICUS, L. Toadstool.

A. rhacodes, Vitt.  
A. portentosus, Fr.  
A. radicatus, Bull.  
A. velutipes, Curt.  
A. dryophyllus, Bull.  
A. sapidus, Kal.

A. cervinus, Schæff.  
A. arvensis, Schæff.  
A. velutinus, P.  
A. campanulatus, L.  
A. fimicola, Fr.

## COPRINUS, Fr.

C. atramentarius, Bull.  
C. picaceus, Bull.

## HYGROPHORUS, Fr.

H. ceraceus, Schæff.

## RUSSULA, Fr.

R. decolorans, Fr.

## MARASMIUS, Fr.

M. prasiosmus, Fr.  
M. rotula, Fr.

## PANUS, Fr.

P. rudis, Fr.

## SCHIZOPHYLLUM, Fr.

S. commune, Fr.

## POLYPORUS, Fr.

P. brumalis, Fr.  P. applanatus, Fr.
P. Boucheanus, Fr.  P. cinnabarinus, Fr.
P. varius, Fr.  P. hirsutus, Fr.
P. lucidus, Fr.  P. versicolor, Fr.
P. resinosus, Fr.

## HYDNUM, L.

H. coralloides, Scop.

## STEREUM, Fr.

S. spadiceum, Fr.

## CLAVARIA, L.

C. cristata, Holm.

## LYCOPERDON, Tourn.

L. giganteum, Batsch.  L. pyriforme, Schæff.

## CYATHUS, Pers.

C. striatus, Hoffm.

## MORCHELLA, Dill.

M. esculenta, Pers.  Morel.

## HELVELLA, L.

H. elastica, Bull.

## PEZIZA, L.

P. coccinea, Jacq.

# SUPPLEMENT

— TO THE —

# FLORA OF THE MIAMI VALLEY,

## For 1878.

---

### SERIES I.

### PHÆNOGAMOUS OR FLOWERING PLANTS.

Acer spicatum, Lam.
Aconitum uncinatum L.    ?
Æsculus flava, Ait.
Alnus serrulata, Ait.
Anagallis arvensis, L.
Anemone nemorosa, L.
Andropogon Virginicus, L.
Anychia dichotoma, Michx.
Apios tuberosa, Mœnch.
Aphyllon uniflorum, Torr. and Gray.
Arabis Drummondii, Gray.    ?
Arabis Ludoviciana, Meyer.
Archemora rigida, DC.
Arenaria stricta, Michx.
Aristida tuberculosa, Nutt.
Aristolochia Serpentaria, L.
Artemisia biennis, Willd.
Aster carneus, Nees.
Aster ericoides, L.
Aster Novæ, Angliæ, L.
     var. roseus, Gray.
Aster oblongifolius, Nutt.

Aster patens, Ait.
Aster prenanthoides, Muhl.
Aster sagittifolius, Willd.
Aster Shortii, Boot.
Astragalus Canadensis, L.
Bidens connata, Muhl.          .
Bidens cernua, L.
Blephilia ciliata, Raf.
Blephilia hirsuta, Benth.
Bœhmeria cylindrica, Willd.
Brassica campestris, L.
Bupleurum rotundifolium, L.    ?
Cacalia suaveolens, L.
Carex Emmonsii, Dew.    ?
Carex lanuginosa, Michx.
Carex umbellata, Schk.    ?
Chenopodium ambrosioides, L.
Chenopodium urbicum, L.
Circæa alpina, L.
Cirsium arvense, Scop.    ?
Collinsonia Canadensis, L.
Convolulus arvensis, L.    ?
Coreopsis tripteris, L.
Cornus alternifolia, L.
Cornus asperifolia, Michx.    ?
Cornus sericea, L.
Cratægus Crus-galli, L.
Cuscuta decora, Chois.
Cuscuta glomerata, Chois.
Cuscuta Gronovii, Willd.
Cynthia Virginica, Don.

Cypripedium candidum, Muhl. ?
Daucus Carota, L.
Desmodium rotundifolium, DC.
Desmodium viridiflorum, Beck
Dioscorea villosa, L.
Dysodia chrysanthemoides, Lag.
Eleocharis acicularis, R. Br.
Epilobium angustifolium, L.
Epilobium coloratum, Muhl.
Erigeron strigosum, Muhl.
Euonymus Americanus, L.
Eupatorium altissimum, L. ?
Eupatorium sessilifolium, L.
Euphorbia corollata, L.
Euphorbia maculata, L.
Euphorbia Peplus, L. ?
Fedia radiata, Michx.
Galium trifidum, L.
Gentiana Andrewsii, Griseb.
Gentiana serrata, Gun.
Gentiana quinqueflora, Lam.
Gerardia sectacea, Walt.
Gerardia quercifolia, Pursh.
Glyceria elongata, Trin. ?
Glyceria nervata, Trin.
Gnaphalium polycephalum, Michx.
Gnaphalium purpureum, L. ?
Gymnostichum Hystrix, Schreb.
Hepatica triloba, Chaix.
Heracleum lanatum, Michx.
Houstonia cærulea, L.

Houstonia purpurea, L.
  var. longifolia, Gray.
Humulus Lupulus, L.
Hypericum mutilum, L.
Hypericum sphærocarpon, Michx.
Ipomœa pandurata, Meyer.
Isopyrum biternatum, Torr and Gray.
Lathyrus palustris, L.
Lemna minor, L.
Lemna polyrrhiza, L.
Lespedeza violacea, Pers.
Liatris spicata, Willd.
Lilium superbum, L.    ?
Liparis liliifolia, Rich.
Lobelia Kalmii, L.    ?
Lobelia spicata, Lam.
Lonicera flava, Sims.
Lonicera parviflora, Lam.    ?
Lophanthus nepetoides, Benth.
Lycopus rubellus, Mœnch.
Lycopus sinuatus, Ell.
Lysimachia lanceolata, Walt.
Lysimachia quadrifolia, L.
Myosotis laxa, Lehm.
Nabalus altissimus, Hook.
Nesæa verticillata, H. B. K.
Ornithogalum umbellatum, L.
Panicum dichotomum, L.
Pedicularis Canadensis, L.
Phaseolus diversifolius, Pers    ?
Phragmites communis, Trin.

Physostegia Virginiana, Benth.
Poa compressa, L.
Polygonum acre, H. B. K.
Polygonum Pennsylvanicum, L.
Polygonum sagittatum, L.
Pyrus arbutifolia, L.
Ranunculus fascicularis, Muhl.
Rosa blanda, Ait.
Rosa micrantha, Smith.
Rosa rubiginosa, L.
Rubus Canadensis, L.
Rubus occidentalis, L.
Rudbeckia speciosa, Wend.
Rumex Brittanica, L.
Rumex verticillatus, L.
Salix longifolia, Muhl.
Salix nigra, Marsh.
Scirpus atrovirens, Muhl.
Scirpus Eriophorum, Michx.
Scirpus lineatus, Michx.
Scirpus pungens, Vahl.
Scutellaria galericulata, L. ?
Silene Armeria, L. ?
Silene noctiflora, L. ?
Silphium laciniatum, L. ?
Silphium terebinthinaceum, L.
Sium lineare, Michx.
Smilacina stellata, Desf.
Solidago stricta, Ait. ?
Sorghum nutans, Gray.
Spartina cynosuroides, Willd.

Spiræa opulifolia, L.
Stachys aspera, Michx.
     var. glabra, Gray.
Stylophorum diphyllum, Nutt.
Thalictrum purpurascens, L.
Thaspium aureum, Nutt.
Thasphium trifoliatum, Gray.
Trillium erectum, L.
Trillium nivale, Ridd.
Verbena officinalis, L.   ?
Vibernum acerifolium, L.
Viola blanda, Willd.
Viola canina, L.
     var. sylvestris, Regel.
Xanthium spinosum, L.
Zizia integerrima, DC.
Zygadenus glaberrimus, Michx.   ?

## SERIES II.

CRYTOGAMOUS OR FLOWERLESS PLANTS.

### FILICES.

Pteris aqilina, L.

### MUSCI.

Barbula convoluta, Hedw.
Bryum cæspiticium, L.
Bryum capillare, Hedw.
Fissidens decipiens.
Hypnum plumosum, Eu. Bry.
Schistidium apocarpum, Br. and Sch.

# HEPATICÆ.

Aneura sinuata,
Asterella hemisphœrica, Beauv.
Calipogeia Trichomanis, Corda.
Davallia rupestris,
Frullania fragilifolia,
Radula complanata, Dum.

# FUNGI.

Agaricus Americanus, Peck.
Agaricus bombycinus, Schæff.
Agaricus campestris, L.
Agaricus cyathiformis, Bull.
Agaricus galericulatus, Scop.
Agaricus hæmatopus, P.
Agaricus infundibuliformis, Schæff.
Agaricus maximus, Fl. Wett.
Agaricus melleus, Fl. D.
Agaricus Morgani, Peck.
Agaricus sinuatus, Fr.
Agaricus strobiliformis, Vitt.
Agaricus ulmarius, Bull.
Agaricus vaginatus, Bull.
Boletus chrysenteron, Bull.
Boletus magnisporus, Frost.
Boletus strobilaceus, Scop.
Cantharellus cibarius, Fr.
Clavaria apiculata, Fr.
Coprinus micaceus, Bull.
Coprinus plicatilis, Curt.
Coprinus radiatus, Bolt.

Cortinarius violaceus, L.
Dædalea unicolor, Bull.
Hygrophorus cossus, Sow.
Lentinus cochleatus, P.
Lentinus Le Contei,
Lentinus vulpinus, Fr.
Lenzites betulina, L.
Marasmius androsaceus, L.
Polyporus abietinus, Dicks.
Polyporus adustus, Willd.
Polyporus frondosus, Fl. D.
Polyporus gilvus, Schwein.
Polyporus graveolens, Schwein.
Polyporus Morgani, Frost.
Polyporus picipes, Fr.
Polyporus radiatus, Sow.
Polyporus sulphureus, Bull.
Russula furcata, P.
Strobilomyces strobilaceus, Berk.
Trogia crispa, P.